40

W.O.

3154

I0191554

TRAINING OF AERIAL MACHINE GUNNERS.

ISSUED BY THE GENERAL STAFF.

JULY, 1916

(Reprinted November, 1916).

(B337)   4000   11/16   H&S   3888wo

TRAINING OF AERIAL MACHINE GUNNERS.

The following notes and practices are an outline of the instructional methods in vogue in the Royal Flying Corps for training machine gunners for aerial work.

The aim of the instruction is to familiarise the gunner with his weapon so that the use of it becomes purely automatic, and therefore no time is lost in opening fire and in remedying any stoppages that may occur.

To that end the Lewis gun—being the arm chiefly used by the Royal Flying Corps—is always handled whenever firing practice is being carried out, even if ·22-inch or other ammunition is being used.

The method of instruction may be briefly outlined as follows :—

(a.) Training in mechanism, stripping, care of the weapon and its appurtenances, and as much ground shooting as possible are done in squadrons, under selected instructors who have passed through a special course of instruction at the Machine-Gun School, Royal Flying Corps.

This course of instruction may be spread over as extended a period as the Officer Commanding squadron may desire, but a minimum of 15 hours is laid down for it.

Upon the conclusion of this course and previous to the departure of the pupil to the Machine-Gun School, a series of tests are applied. If the pupil does not obtain a necessary percentage of marks he receives further instruction in the squadron.

These tests are included in this pamphlet, and also a syllabus showing allotment of hours of instruction.

(b.) All aerial practices and advanced instruction in the Lewis and Vickers guns are carried out at the Machine-Gun School, Royal Flying Corps.

The School is thus relieved from giving elementary instruction in the Lewis gun and can devote its whole attention to aerial practice as distinct from ground practice.

Upon arrival at the School the pupils are tested by the Staff with a view to grouping the pupils in squads according to capabilities and knowledge.

If a pupil possesses an insufficient knowledge of the Lewis gun he is returned to his unit.

This is necessary in order to enable only advanced instruction to be given at the School.

The aerial course lasts from two to three weeks, being naturally dependent upon weather conditions.

Two oral examinations are held during the course in all subjects dealt with, and a report as to each pupil's capabilities, &c., is rendered direct to the Director of Air Organization.

SQUADRON TRAINING.

The syllabus attached has been drawn up with a view to giving officers and non-commissioned officers who are charged with instruction in the Lewis machine gun an idea of the amount of time needed in various subjects in a short squadron course of 12 working days.

The object is to give all pupils as thorough a working knowledge of the mechanism, care and handling of the gun, as is possible in so short a space of time.

The instructor, before commencing to teach exact details, such as names of parts, &c., should briefly explain the general principle of the gun and give a rough idea of the working of the various parts.

He should illustrate, if possible, the uses of various parts by apt similes in order to more readily fix the information in the pupils' minds.

All instruction should be made as interesting as possible.

It is essential that the attention of classes should be held throughout the lectures, and no instruction on any one subject should be carried on for so long as to become boring.

Before proceeding with a further lesson on any particular subject the pupil should be catechised on the previous lesson in order to ascertain if any doubt exists upon any particular point.

As instruction advances it will be found a good point to detail pupils to act as instructors in various items.

This will enable the instructor to ascertain how much knowledge has been assimilated by the pupils and tends to give the pupils confidence.

Pupils should also be asked to catechise upon various subjects which have been dealt with.

Classes should be made as comfortable as possible when listening to explanations, but the instructor should occasionally interpose a question on the subject being explained in order to ascertain if attention is being paid to the instruction.

Squads, if possible, should be limited to six members, as individual instruction in mechanism is difficult and almost impossible with a larger number.

Equipment necessary for each squad—

1 Tripod for Lewis gun, obtainable from O.A.D., Farnborough.

2 Lewis guns, one to be used for mechanism, stripping and drill, and one for shooting. Complete with spares.

20 Dummy cartridges. (These can be made from ball ammunition if desired, but holes *must* be pierced in them to show they are dummies.)

1 Deflector and bag, obtainable from O.A.D., Farnborough.

1 Mounting yoke, obtainable from O.A.D., Farnborough.

1 Fixed sight and barleycorn foresight, obtainable from O.A.D., Farnborough.

A supply of flannelette and lubricating oil for use with the guns,

SYLLABUS OF SQUADRON COURSE.

12 Working Days.

First Day.	Fifth Day.	Ninth Day.
Drill, 1 hour. General description of gun, 1 hour. Mechanism, 1 hour.	Mechanism and stripping, 1 hour. Care of guns, 1 hour. Mechanism, 1 hour.	25 yards range. Part I.*
Second Day.	**Sixth Day.**	**Tenth Day.**
Drill, ½ hour. Mechanism, 1½ hours. Drill, 1 hour; and immediate action.	Drill, 1 hour. Mechanism and stripping, 1 hour.	250 yards range. Part II.*
Third Day.	**Seventh Day.**	**Eleventh Day.**
Mechanism and stripping, 1½ hours. Drill, ½ hour. Mechanism, 1 hour.	Resumé of work done, 1 hour. Drill and stoppages, 1 hour. Mechanism and stripping, 1 hour.	Stoppages on 25 yards range.*
Fourth Day.	**Eighth Day.**	**Twelfth Day.**
Mechanism and stripping, 1 hour. Care of guns, 1 hour. Drill, ½ hour. Catechism and drill, ½ hour.	Description of course to be fired, 1 hour. Drill and rehearsal of practices, 1 hour. Standard tests, 1 hour.	Examination as to fitness for Advanced Course at Machine-Gun School.

* Pupils are to prepare guns for firing and also clean them upon conclusion of practices.

Total Time allotted.

	hours.
Drill	7
Mechanism and stripping	11
Lectures	2
Care of guns	2
Shooting time, uncertain	—
Standard tests	1
	23

STANDARD TESTS IN MACHINE-GUN TRAINING FOR ROYAL FLYING CORPS.

These tests have been drawn up with a view to enabling an instructor to determine if a pupil is fit at the end of a course in a squadron to undergo an advanced course in aerial firing at the Machine-Gun School, Royal Flying Corps.

75 per cent. of marks in each subject should be obtained before a pupil is considered satisfactory.

TESTS.

1. *Prepare gun for firing.*—(*a.*) Oiling. (*b.*) Assembling, this includes the points to be attended to before taking a gun up in a machine. (*c.*) Testing ammunition. (*d.*) Testing magazines. For this test all available gun spares should be upon the table.

Instructor to deduct marks at his discretion.

(15 marks.)

2. *Tests of stripping*—

(*a.*) *Change cartridge guide spring.*

Time allowed, 10 seconds.

Gun not to be stripped in this test.

A dummy cartridge and spare cartridge guide spring should be by the pupil.

Gun to be mounted on a tripod and pupil seated as for firing.

(*b.*) *Change 1 extractor.*

Time allowed, 1 minute 20 seconds.

Spring must be at correct tension, from 11 to 12 lbs., upon conclusion of this test,

A dummy cartridge and spare extractor should be by the pupil.

Gun on table.

(*c.*) *Change return spring.*

Time allowed, 1 minute 30 seconds.

Spring must be at correct tension upon conclusion of this test.

A dummy cartridge and spare return spring, with spring tension lowered, should be near pupil.

Gun on table.

(*d.*) *Change all above.*

Time allowed, 2 minutes.

A dummy cartridge and spares necessary to be near pupil.

Gun on table.

(20 marks for whole of these tests (*a.*), (*b.*), (*c.*), (*d.*).)

Points to be deducted.—One point for each 5 seconds over-time in each of (*a.*), (*b.*), (*c.*), or (*d.*). One point for every fault discovered when pupil concludes each of (*a.*), (*b.*), (*c.*), or (*d.*).

3. *Tests of drill in instructional room.*— Changing six magazines.

Six empty magazines to be by side of gun.

Gun on tripod, pupil seated as for firing.

Magazines to be fitted to gun one by one, gun to be cocked, aim to be taken, trigger to be pressed, immediate action to be gone through after pressure of trigger (without taking eyes off mark), and magazines to be removed.

This to be repeated until the six magazines have been gone through.

Time allowed, 1 minute.

(25 marks.)

Points to be noted by pupil.—(*a.*) Stream lining of magazine when fitting it or removing it.

(*b.*) Noting if magazine is on correctly.

(*c.*) Immediate action with eye on object.

(*d.*) Unloading at end of test, included in time limit.

Deduction of marks.—One mark to be deducted for each error committed under (*a.*), (*b.*), (*c.*) and (*d.*).

4. *Drill on range.*—(*a.*) As for 3, but if a ·22-inch Winchester automatic rifle can be obtained it should be fitted to gun so that the trigger of the Lewis gun when pressed will release trigger of ·22-inch rifle.

Both guns should be harmonized so that aim taken with the gun-sights directs the ·22-inch rifle on same mark. Magazine of ·22-inch rifle to have six rounds in for this test.　　(25 marks.)

Procedure as in 3, but result of aiming to be seen practically by strike of shot on target at 10 yards. Marks to be deducted as in 3, and also 1 mark to be deducted for every inch by which the group exceeds 4 inches in diameter.

(*b.*) If no ·22-inch rifle is obtainable, the same tests can be carried out at 25 yards with ball ammunition, as follows:—

> Six magazines, each containing one round, so that it will be fed in when gun is cocked.

Size of group, 4 inches.

Deduct points as in (*a*), but two marks to be deducted for every inch above 4 inches.

Time allowed in (*a*) or (*b*), 1 minute 15 seconds.

5. *Mechanism.*—Pupil should be able to give an explanation of the working of any part of the mechanism of the gun.

Marks to be deducted at discretion of instructor.

(10 marks.)

For this test, gun and any necessary spares, such as piston rods, bolt, &c., to be on table beside gun. Also a magazine and dummies.

6. *Shooting.*—Instructional practice at 25 yards does not come into tests.

Classification practice, 250 yards.—10 rounds ranging at target 10 feet by 3 feet.

100 rounds application.

Five magazines, each containing 20 rounds, to be fired at three targets, 10 feet by 3 feet. placed side by side.

Each target to have vertical spaces marked on them, 20 inches apart.

Firer to have loaded and to be aiming.

Fire to commence on command " Fire."

Targets to be "searched" from one end to the other, in bursts of 10 rounds.

Time limit, 1 minute 40 seconds; but firer is not to be told if he is exceeding this limit.

Score to be signalled—

       Total number of hits.

       Spaces missed.   (15 marks.)

Deduction.—Not more than 5 spaces should be missed. Two marks to be deducted for every space missed. One mark to be deducted for every 5 seconds beyond time limit.

As all squadrons are not able to obtain a 250 yards range this practice is not always possible and may be carried out at Machine-Gun School.

7. *Stoppages on* 10 *or* 25 *yards range.*—Three magazines, each containing 10 rounds. Two stoppages arranged in each magazine.   (20 marks.)

Deductions.—One point to be deducted for each error in application of immediate action or remedy.

NOTE.—It is not intended that any of these tests should be taken as a test of the efficiency of a pupil for duty as a gunner overseas. They are merely designed to ascertain, as far as possible, in squadrons the extent of a pupil's knowledge of the mechanism and care of the gun, and his proficiency in handling it preparatory to proceeding to the Machine-Gun School for an advanced course.

GUNS, APPLIANCES, TARGETS, &c., AVAILABLE AT THE
MACHINE-GUN SCHOOL, ROYAL FLYING CORPS, FOR
TRAINING, AND THE LESSONS TO BE TAUGHT WITH
EACH.

(*a.*) *·22-inch Winchester automatic rifle.*—This is attached to
the Lewis gun and is so adjusted that the line of sight of the
Lewis gun coincides with the strike of the bullet of the ·22-inch
automatic at 25 or 50 yards, as desired.

The triggers of each are connected so that when the Lewis
gun trigger is pressed the ·22-inch automatic fires.

In the earlier practice, the rifle is loaded with 10 rounds, the
range is 10 yards, and all shots must be in a 3-inch ring.

The instructor checks pupil for incorrect holding, &c.

The object of this practice is to ascertain the aiming and
trigger-pressing power of the pupil.

In the later practice the pupil is required to execute the drill
necessary to keep the gun in action during an aerial combat, *i.e.*,
loading, firing, testing magazine to see if it is empty, or if a
stoppage has occurred, changing magazine and relaying.

He fires one round from the Winchester ·22-inch each time
the trigger of the Lewis gun is pressed—this is taken to be one
complete magazine fired—and is required to group his shots in
a 3-inch ring at 10 yards, and to execute his drill correctly;
otherwise marks are deducted.

Ten magazines are to be fitted to the gun and ten rounds are
in the Winchester rifle.

A further practice is carried out with this attachment at a
moving model aeroplane at 25 yards.

The aeroplane is suspended from a wire and moves down-
wards at a fair speed and across a front of 10 yards.

The gun is loaded with 10 rounds and the firer fires one
round on each run.

The firer is here taught the necessity for quick working
of eye, brain and finger which is so essential in aerial
gunnery.

The pupil is also trained at small free balloons with this
apparatus with the object of stimulating his interest, as the
target bursts when it is struck.

(*b.*) *Browning automatic shot-gun.*—This is attached as in (*a*).

The gun takes five rounds in the magazine and the pupil fires
at free balloons whilst on the ground, and later at free balloons
when in an aeroplane.

The advantage of this apparatus is that a pupil can follow
balloons over any area and can fire at them without danger to
the inhabitants.

He is required to fit one magazine between each shot as
though firing the actual Lewis gun.

This practice stimulates interest, as the target disappears
when hit.

The pilot is required to observe as far as possible the correctness or otherwise of the pupil's actions.

(c.) *Gun camera.*—A special camera is attached to the Lewis gun, and the trigger is attached to the shutter release of the camera, so that when the trigger is pressed the film is exposed.

The line of sight of the gun coincides with the field of view of the camera.

Two pupils are sent up, each in an aeroplane, and each having a camera loaded with one roll of films.

They are told to attack one another and open fire as if in attack.

The films are developed and later such faults as incorrect aiming, deflection, position of attack, and holding of gun are pointed out by the instructor.

The camera has a screen which is marked off in squares in order to show errors in deflection.

This test is exceedingly useful as it practically shows all faults which are committed in the air in aiming, aiming off and holding.

By faulty holding is meant the want of proper check in the movement of the gun during firing.

The gun is not fired in this practice, but the vibration of the machine will cause a blurred image if the gun is not held rigidly.

(d.) *Lewis machine gun.*—The mechanism, stripping, stoppages and drill, together with precautions necessary for safety in the air, are taught.

A graded series of land and aerial practices are fired and culminate in firing at a target towed by a tractor aeroplane. This latter firing is carried out with ordinary battle sight, special deflection and automatic sights.

Practice in remedying stoppages is also given in the air.

(e.) *Vickers gun.*—As far as possible sufficient instruction is given in this gun to enable pilots to use it in single seaters or in future tractors, and remedy stoppages which may occur.

(f.) *Training for gunners in 2-seater machines.*— A Vickers' nacelle is fitted as for flying, and all drill and land practices are fired from it in order to accustom firers to cramped positions.

(g.) *Training for pilots in single-seaters* —A dummy nacelle is provided. It is fitted with rudder and elevator controls, which will move the nacelle horizontally or vertically, and also with a gun and magazine boxes.

The instructor can move the nacelle at will, and the pilot has to manœuvre the gun or nacelle back into position by means of controls.

Drill and land practice is carried out from this dummy.

(h.) *Deflection teacher.*—A Le Prieur deflection teacher is used. It shows the deflection necessary when firing from the side of own machine—*i.e.*, at right angles to line of flight— at an enemy machine on any angle of flight.

(*j.*) *Le Prieur sight or corrector.*—This is taught and used practically on the ground. Fire is at fixed targets, and the pupils adjust the sight for any supposed angle and speed of enemy machine.

The correct position of shots is known for desired angles of flight and speeds, and pupils are checked accordingly.

This practice is first carried out slowly, and later rapidly in order to instil into pupils confidence in the use of the sight.

A device is used by means of which a pupil's powers of accurately and quickly determining the enemy's angle of flight is tested and developed in order to be able to use the Le Prieur corrector successfully.

This instruction is essential as the angle of flight is very difficult to determine correctly.

Finally, this sight is used in a Vickers F.B. against a balloon towed by a tractor aeroplane at a speed of about 60—65 miles per hour. This is the most difficult practice outside actual war conditions.

TARGETS USED.

A.—25 *yards target.*— A canvas target, 4 feet by 2 feet, faced with paper.

Upon this are pasted a 3-inch black square for early grouping practices and five black strips, 1 inch by 2 inches, for searching practices.

B.—250 *yards targets.*—The usual brown screens used on rifle ranges for machine-gun training are utilized.

The screens are 10 feet by 3 feet, and are of canvas faced with brown paper.

As many as necessary may be placed side by side to form a long target.

The screens are each divided into 20-inch bands by ruling with blue pencil.

This is done in order to insure searching fire being used by pupils, as results are judged by the number of hits obtained and the number of spaces missed.

C.—*Model aeroplane for miniature range.*—A model aeroplane of tin is used; it is suspended on a wire along which it moves on small pulley wheels.

A fair rate of speed is obtained and the pupils' training in quickness of aim and rapidity of opening fire is carried on at this target.

D.—*Balloons.*—Rubber balloons of various sizes are filled with hydrogen and are used in various practices both stationary and free.

The early practices are fired at balloons which are attached by strings to a wooden framework, but are given a certain amount of swaying movement by the wind. The pupils' interest is aroused inasmuch as the balloons burst upon being hit.

Quick decision in aiming and pressing the trigger is taught in this.

In later practices balloons are allowed to go free and the pupil fires at them from the ground and then from an aeroplane.

In firing from the aeroplane the pupil is required to direct the pilot to manœuvre the machine in order to get into the best position for opening fire.

Thus the pupil is taught the necessity for close co-operation between pilot and gunner in aerial fighting.

E.—*Target for early aerial practice.*— A canvas target representing an aeroplane is placed at an angle upon the shingle.

The firer—first in a pusher machine and later in a tractor—is flown over the target at a height of not more than 500 feet.

He opens fire and observes the strike of his shots on the shingle, and applies his fire in order to strike the target upon subsequent runs.

The firer in these practices is shown the necessity for quick decision in opening fire and the safety precautions necessary when firing in the air.

F.—*Target towed by tractor aeroplane.*—A hollow canvas cone, stiffened with wood, is towed behind a B.E. 2c. The cone has attached to it a net which contains a large, hydrogen-filled, rubber balloon; this protrudes from the larger end of the cone.

The whole is towed about 200 feet behind the aeroplane.

The speed of the aeroplane is very little reduced by this towing, and the cone can be drawn in before landing by means of a reel or winch near the pilot's right hand.

The firer is first in a Vickers F.B. and later in a B.E. 2c, and various practices are fired at the balloon, attacking it from different angles.

With this target the firer gets the impression of firing as in actual combat in the air, and when the balloon is struck it disappears.

PRACTICES TO BE FIRED AT MACHINE-GUN SCHOOL, ROYAL FLYING CORPS.

FROM GROUND.

No.	Practice.	Range.	Rounds.
1	Grouping with ·22-inch Winchester attached to Lewis gun. 10 magazines to be fitted .. To be repeated until satisfactory group of 4 inches is obtained. First without time limit, last 2 shoots with time limit of 2 minutes 20 seconds.	yards. 25	30 (or more if necessary).
2	Grouping ·303 Lewis gun.—As for 1 . ..	25	30 (or more).
3	250 yards practice for squads which have not fired this practice in Squadron Training ..	250	110
4	Fixed balloons with ·22-inch Winchester attached to Lewis gun. 2 sighting shots at target	75	12
5	Free balloons.—As for 4	Unknown	10
6	Fixed balloons, ·303 Lewis gun	200	47 5 and 5 tracer.
7	Free balloons, ·303 Lewis gun	Unknown	47 5 and 5 tracer.
8	Free balloons, Browning automatic shot gun attached to Lewis gun	Unknown	5
9	Model aeroplane moving, ·22-inch Winchester attached to Lewis gun. 6 practices each of 10 rounds	25	60
10	Practice from Vickers' F.B. at 1 sheet on ground. 5 magazines of 10 rounds each ..	yards. ..	50
11	Practice from Vickers' F.B. at 2 sheets. 3 magazines of 20 rounds each	..	60
12	Repeat 10 from B.E. Strange mounting ..	..	50
13	Repeat 11 from B.E. Strange mounting ..	..	60
14	Repeat 10 from B.E. Side mounting ..	..	50
15	Repeat 11 from B.E. Side mounting.. ..	..	60
16	Free balloons, anywhere over aerodrome. Browning automatic shot gun attached to Lewis gun	..	5
17	Fixed balloons, Vickers. 2 magazines ..	..	94 5 and 5 tracer.
18	Fixed balloons, B.E. Strange mounting ..	..	94 5 and 5 tracer.
19	Free balloons, Vickers	..	94 5 and 5 tracer.
20	Free balloons, B.E.	..	94 5 and 5 tracer.

www.ingramcontent.com/pod-product-compliance
Lightning Source LLC
Chambersburg PA
CBHW050842040426
42339CB00014B/84